CW00455040

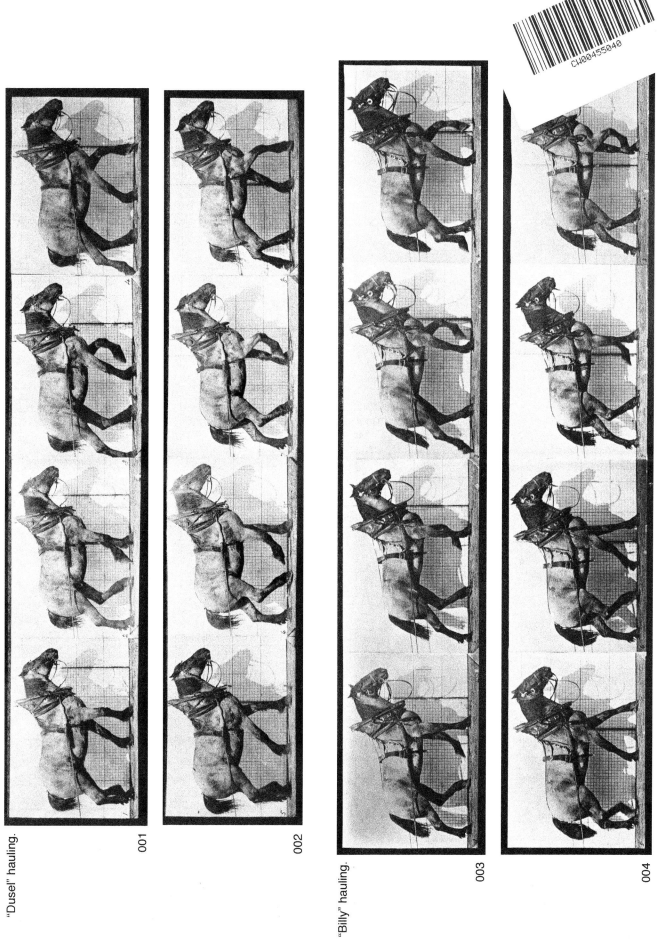

"Dusel" hauling.

001

002

"Billy" hauling.

003

004

1 HORSES

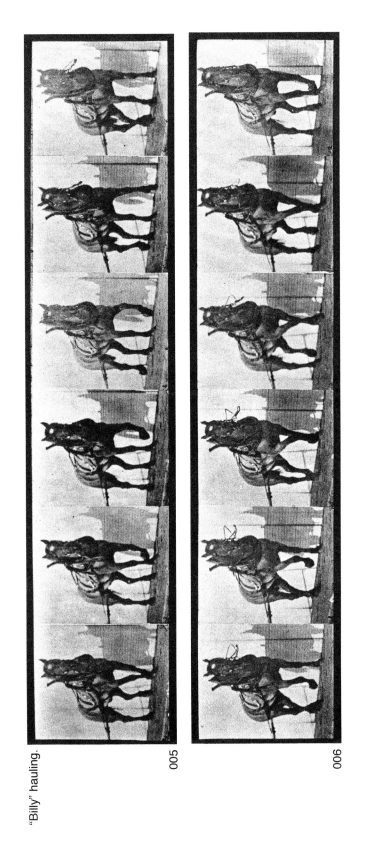

"Billy" hauling.

005

006

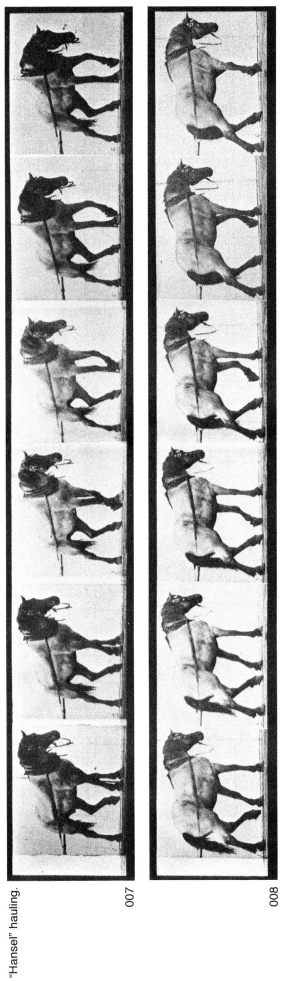

"Hansel" hauling.

007

008

"Johnson" hauling, man pulling at head.

009

010

"Hansel" walking, free.

011

012

"Eagle" walking, free.

013

014

015 "Clinton" walking, mounted, irregular.

4 HORSES

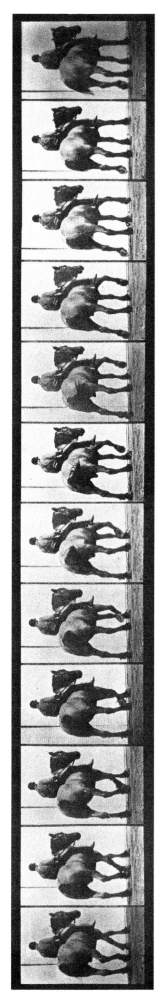

016 "Dusel" walking, bareback.

"Annie G." walking, saddled.

017

018

5 HORSES

"Smith" walking, bareback; rider nude.

019

020

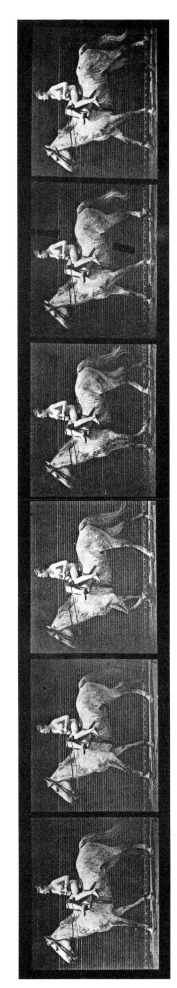

021 "Tom" walking, saddled; female rider nude.

022 "Beauty" walking, saddled, irregular.

"Clinton" ambling, bareback; rider nude.

023

024

025 "Katydid" walking, harnessed to sulky.

7 HORSES

"Pronto" pacing, saddled.

026

027

"Pronto" pacing, harnessed to sulky

028

029

030 "Eagle" trotting, free.

"Daisy" trotting, saddled.

031

032

9 HORSES

"Dusel" trotting, bareback.

033

034

"Beauty" trotting, saddled.

035

036

10 HORSES

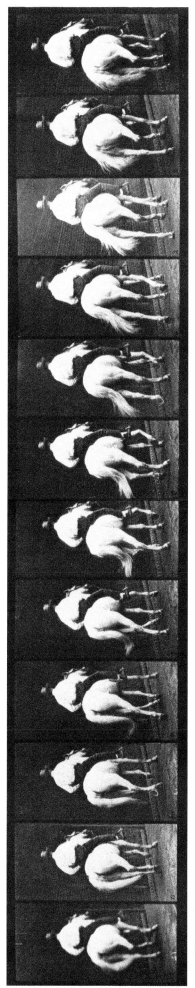

037 "Elberon" trotting, saddled.

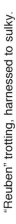

"Reuben" trotting, harnessed to sulky.

038

039

11 HORSES

"Katydid" trotting, harnessed to sulky.

040

041

"Nellie Rose" trotting, harnessed to sulky.

042

043

"Daisy" cantering, saddled.

044

045

"Annie G." cantering, saddled.

046

047

13 HORSES

"Daisy" galloping, saddled.

048

049

"Bouquet" galloping.

050

051

14 HORSES

"Hansel" galloping, bareback.

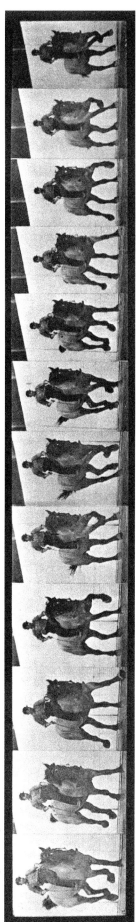

052

053

"Bouquet" galloping, saddled.

054

055

15 HORSES

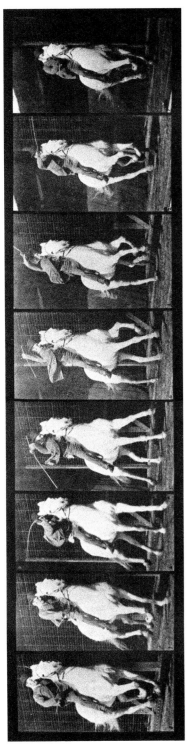

056 "Dan" galloping, saddled.

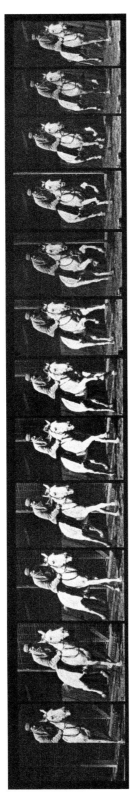

057 "Pandora" galloping, saddled.

"Daisy" jumping a hurdle, saddled, preparing for the leap.

058

059

"Daisy" jumping a hurdle, saddled, clearing, landing, and recovering.

060

061

062

"Daisy" jumping a hurdle, saddled.

063

064

"Pandora" jumping a hurdle, bareback, clearing, and landing; rider nude.

065

066

"Hornet" jumping over three horses.

067

068

069

070

071 Horse rearing.

"Gazelle" walking, spavin, right hind leg.

072

073

"Ruth" bucking and kicking.

074

075

"Ruth," refractory.

076

077

"Denver," refractory.

078

079

"Jennie" walking, bareback; a boy riding.

080

081

082 "Zoo" walking, saddled; a girl riding.

Sow walking.

083

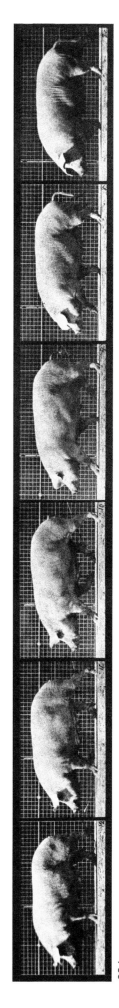

084

23 DOMESTIC ANIMALS

Ox walking.

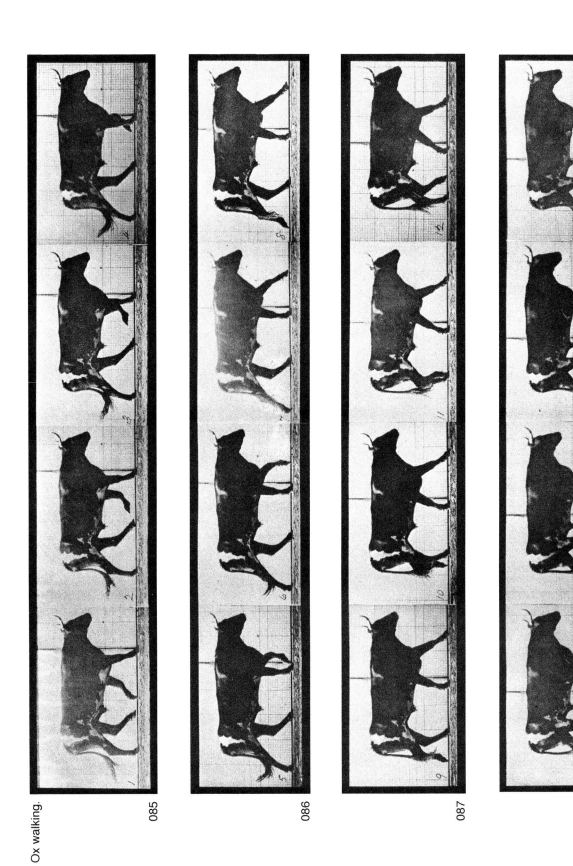

085

086

087

Goat walking.

089

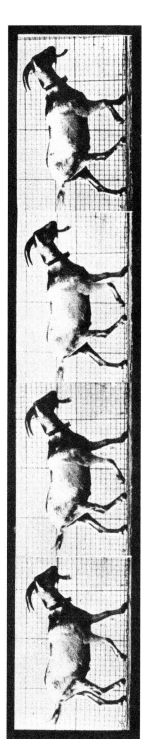

090

Goat trotting, harnessed to a sulky.

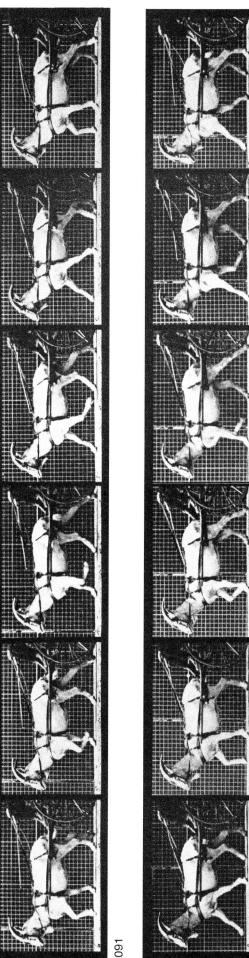

091

092

"Dread" walking, interrupted.

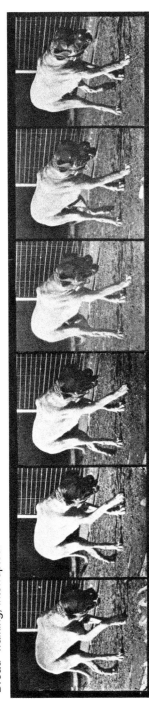

093

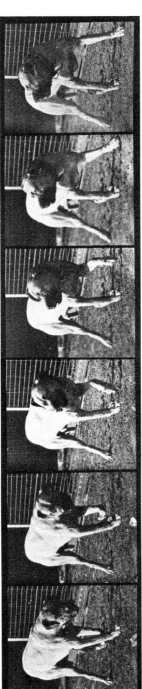

094

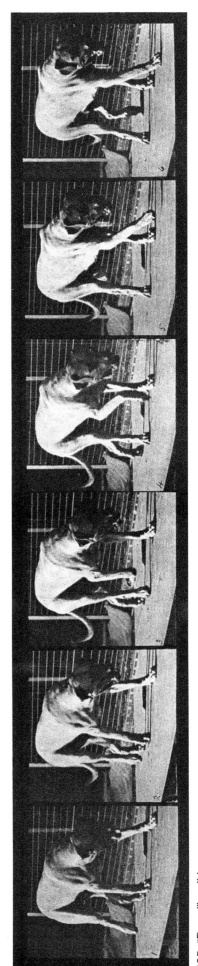

095 "Dread" walking.

"Dread" galloping.

096

097

098

27 DOMESTIC ANIMALS

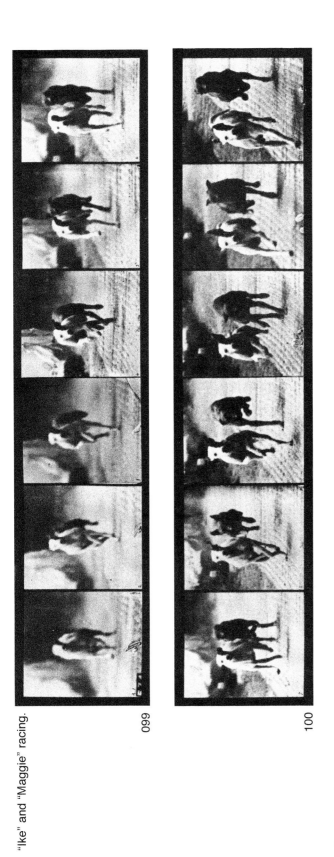

"Ike" and "Maggie" racing.

099

100

"Kate" turning around.

101

102

"Smith" aroused by a torpedo.

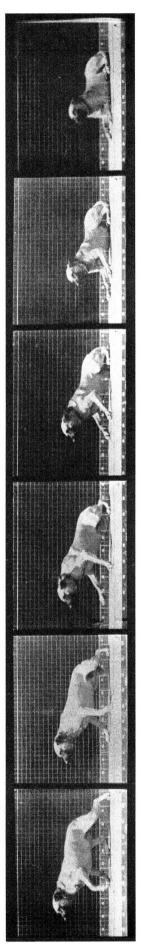

103

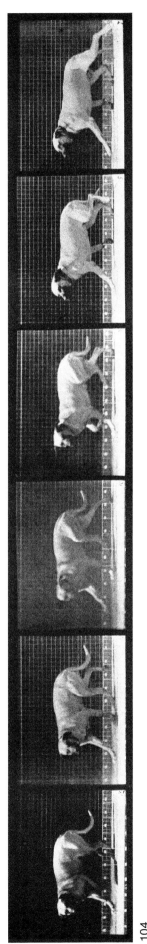

104

Cat galloping.

105

106

29 DOMESTIC ANIMALS

Cat trotting, changing to a gallop.

107

108

109

110

30 DOMESTIC ANIMALS

Fallow deer, buck, trotting.

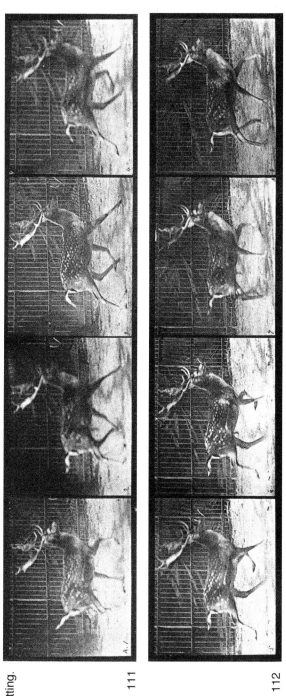

111

112

Fallow deer, buck and doe, galloping.

113

114

31 DOMESTIC ANIMALS

Fallow deer, doe galloping and kid jumping.

115

116

117

Elk trotting.

118

119

Elk galloping.

120

121

Buffalo galloping.

122

123

124

125

Lion walking.

126

127

128

Lion walking and turning around.

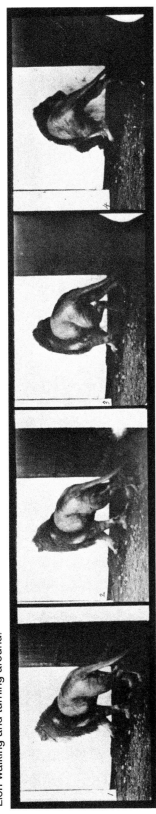

129

130

131

Tigress walking.

132

133

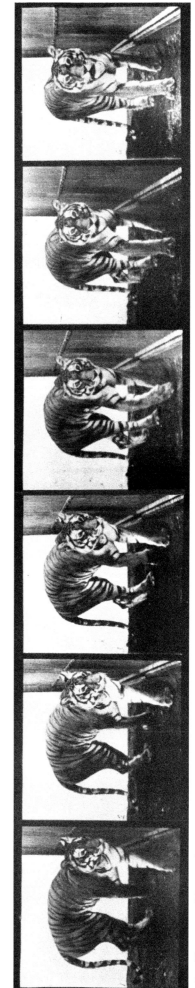

134 Tigress walking and turning around.

Elephant walking.

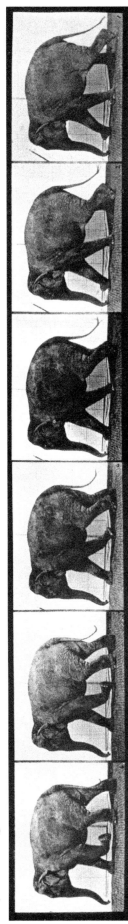

135

136

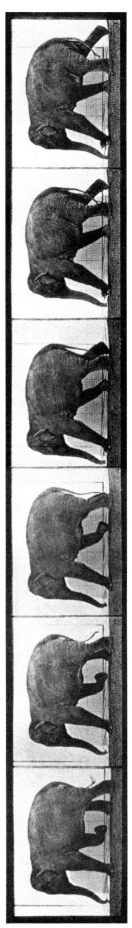

137

Two elephants walking.

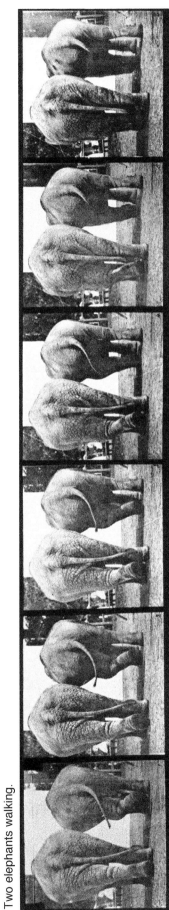

138

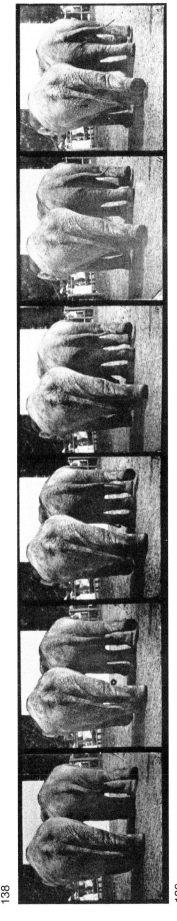

139

Egyptian camel racking.

140

141

142 Bactrian camel galloping.

Raccoon walking, changing to a gallop.

143

144

Baboon walking on all fours.

145

146

147

148

Sloth walking suspended on a horizontal pole.

149

150

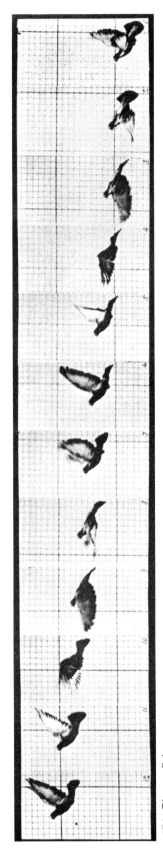

151 Pigeon flying.

Cockatoo flying.

152

153

154

43 WILD ANIMALS & BIRDS

155 Vulture flying.

Cockatoo flying.

156

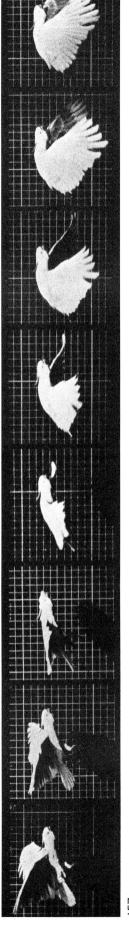

157

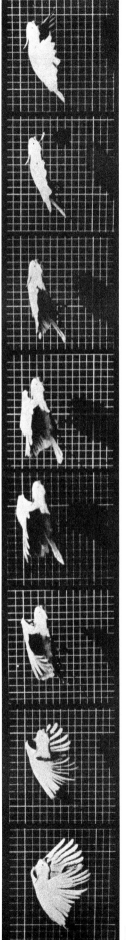

158

44 Wild Animals & Birds

159 Fish hawk flying

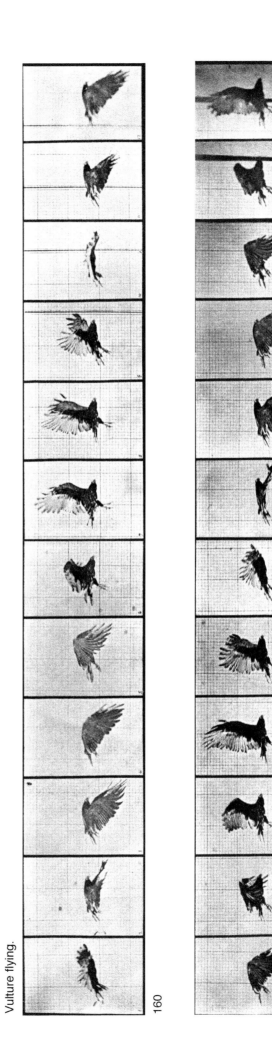

Vulture flying.

160

161

Ostrich running.

162

163

164 American eagle flying.

46 WILD ANIMALS & BIRDS

American eagle flying near the ground.

165

166

167

47 WILD ANIMALS & BIRDS